To Bailey, Aliea, and Audrey, you've brought so much love and laughter into my life. I cherish every moment we've shared and look forward to all the new stories and adventures yet to come. And Michael, thank you for being the rock that anchors us in love and turns every day into a cherished adventure together. ♥

Petoskey Stone Explorers Discover, Learn, Create
First Edition

Published by: Petoskey Press
Printed in the United States of America
ISBN: 978-1-961144-02-6

Author and Illustrator: Michelle Bakerson, Ph.D.
All images taken by Michelle Bakerson, Ph.D.

Michelle Bakerson, Ph.D., is a former kindergarten teacher and tenured professor who spent 16 years training future educators. A lifelong learner and passionate educator, Michelle believes discovery is one of the greatest joys we can give children—and she knows that feeling firsthand.

She began hunting for Petoskey stones with her daughter, often asking, "Is this one?" only to hear, "Not yet." After two years of searching, she finally found her first, and her daughter was right there with her. That unforgettable moment of discovery inspired not only this book, but a renewed love for learning through wonder, patience, and shared joy. Michelle is the founder of Game On Teaching where she creates resources that spark curiosity, build confidence, and make learning an adventure.♥

Author's Note
Some historical details in this story were inspired by information from NorthernMichiganHistory.com: "Honoring Chief Ignatius Petoskey."
https://northernmichiganhistory.com/honoring-chief-ignatius-petoskey

For more information or additional resources contact: gameonteaching@gmail.com

PETOSKEY STONE EXPLORERS

DISCOVER, LEARN & CREATE

Michelle Bakerson, Ph.D.

Have you ever found a rock that looks like it has tiny flowers inside? That might be a Petoskey stone, Michigan's very own state stone!

Real image of a Petoskey stone.

Long ago, in 1787, a baby was born by a river in Michigan. As the sun rose, its rays shined brightly on his tiny face.

His father smiled and said, "He will be named Petosegay, Sunbeam of Promise!"

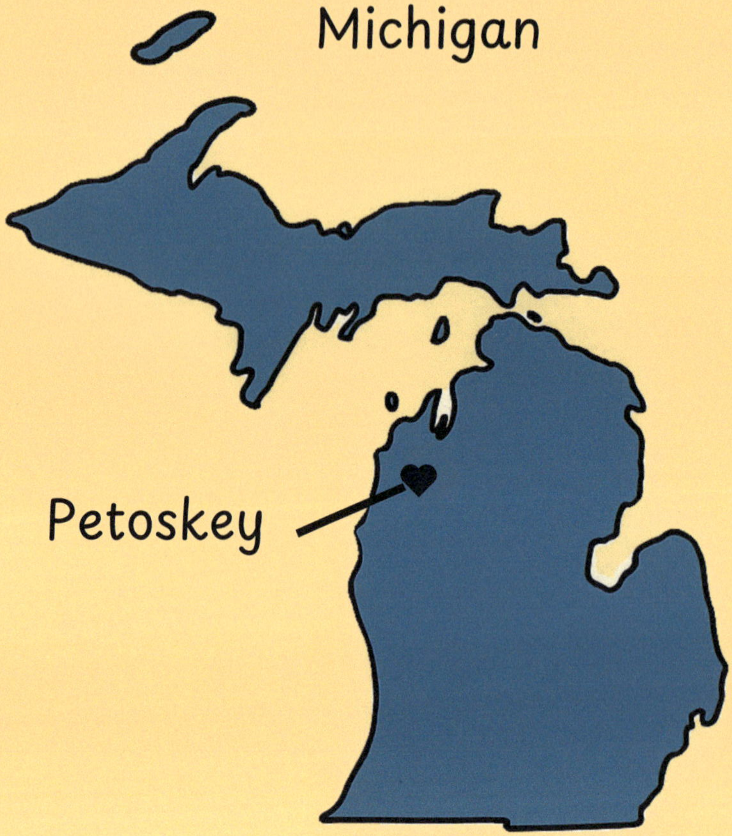

That little baby grew up to be a kind and wise leader. He helped lots of people and lived on the land where a town would one day bloom. They named the town Petoskey, to remember him with love!

And guess what?
That's where Petoskey stones were first discovered! People found them on the beach, along rivers and in gravel pits.

So, what is a Petoskey stone?

It's not just a pretty rock, it's actually a fossil of a coral that lived over 350 million years ago!

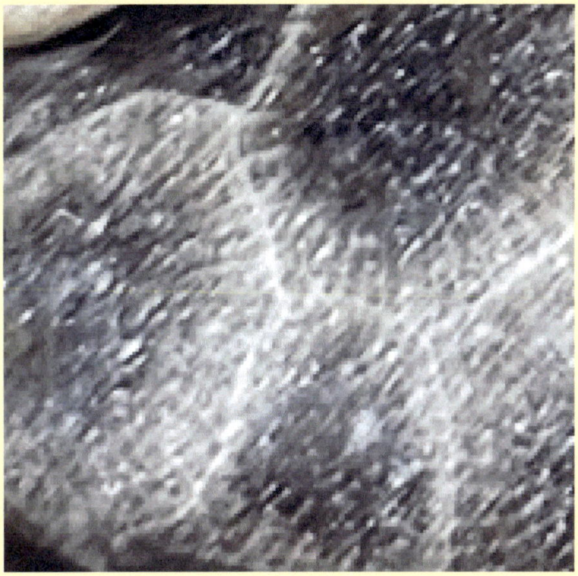

A long, long
time ago,

Michigan wasn't land at all, it was a warm, shallow sea, just like a tropical reef!

Did you know that coral reefs, those big, colorful underwater cities, are made by tiny animals like me?

I'm a polyp, small, soft, and shaped kind of like a little tube. I live in a cozy round home called a corallite. It's like a tiny rock cup that keeps me safe and snug!

And guess what? I have waving arms at the top of my head called tentacles! I use them to grab my food. I love to snack on teeny-tiny floating plants and animals called plankton. Sometimes, I even catch a tiny fish, yum!

But that's not all. I'm also a builder! At the bottom of me, I make a special kind of rock called calcium carbonate.

It's super strong, and when all my polyp friends and I build together, we create something amazing...

...a coral reef! I'm never alone, nope!

I live with lots of polyp pals, all snuggled together. We're connected by a stretchy, living skin called the coenosarc (you say it like "SEE-no-sark").

It's like a big cozy blanket that covers us all.

But one day everything changed.

A long, long time ago, the weather started to get colder and colder.

Michigan wasn't a warm, shallow sea anymore, it became icy and chilly. My coral friends and I couldn't live there any longer.

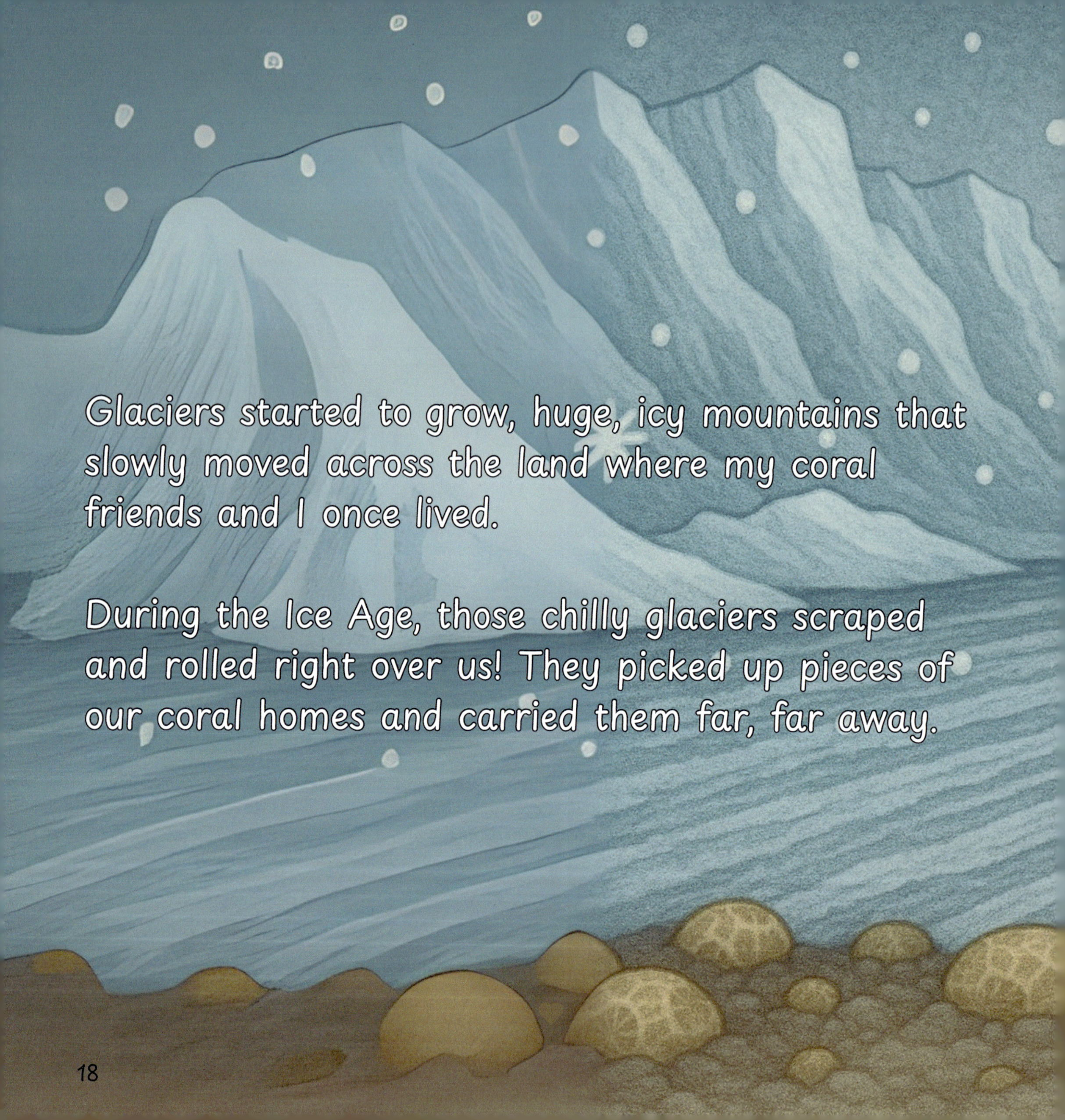

Glaciers started to grow, huge, icy mountains that slowly moved across the land where my coral friends and I once lived.

During the Ice Age, those chilly glaciers scraped and rolled right over us! They picked up pieces of our coral homes and carried them far, far away.

When the ice melted, it left us scattered across Michigan miles from where we first began!

These days, the best place to find me is along beaches, rivers, or even in gravel pits. Your best chance is right after a big storm—when I've been washed up on the shore, just waiting for you to find me!

If you look closely at a Petoskey stone, You'll see tiny shapes like honeycombs or starbursts. Each one was a little coral creature's home!

Top View

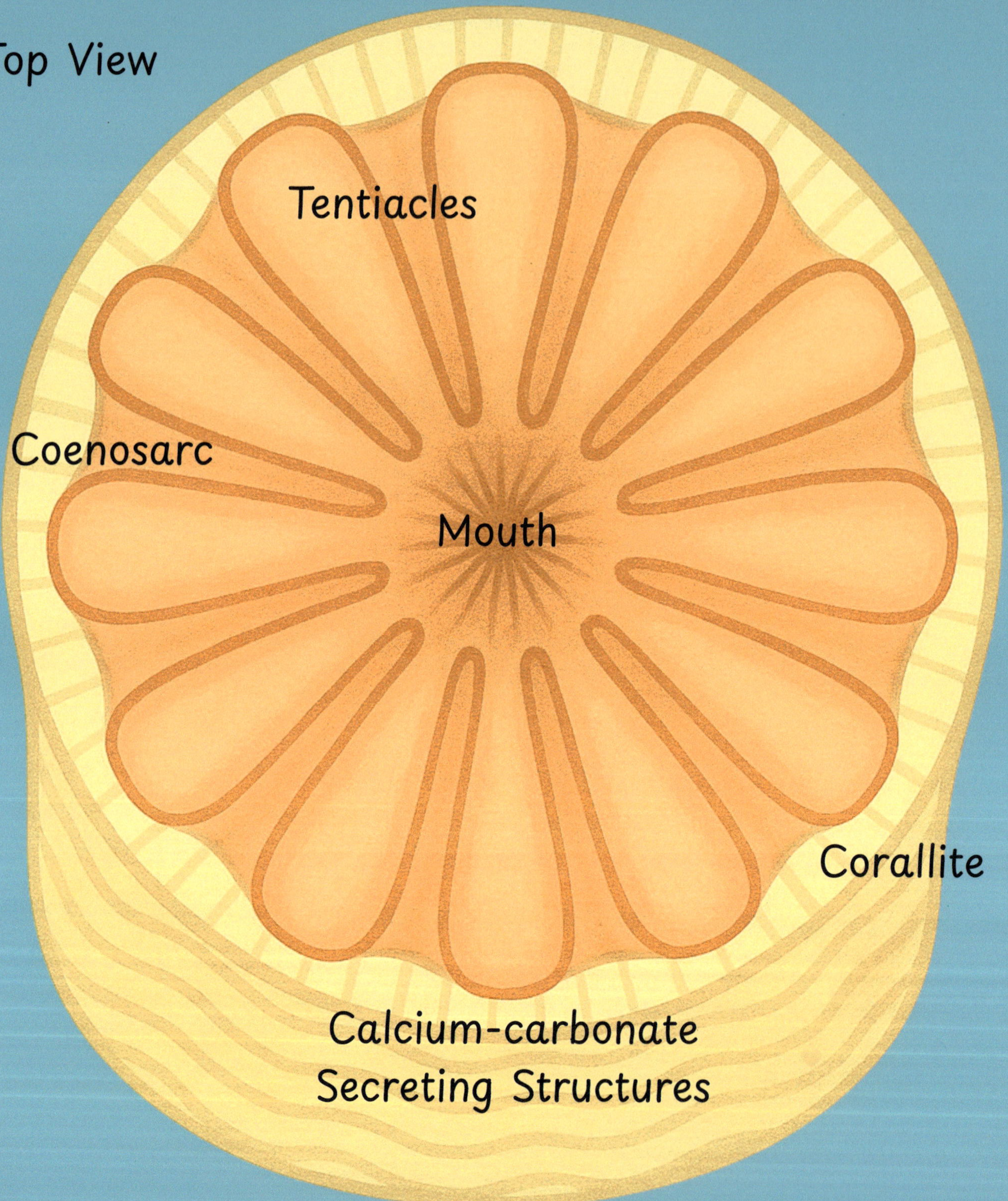
Tentiacles
Coenosarc
Mouth
Corallite
Calcium-carbonate
Secreting Structures

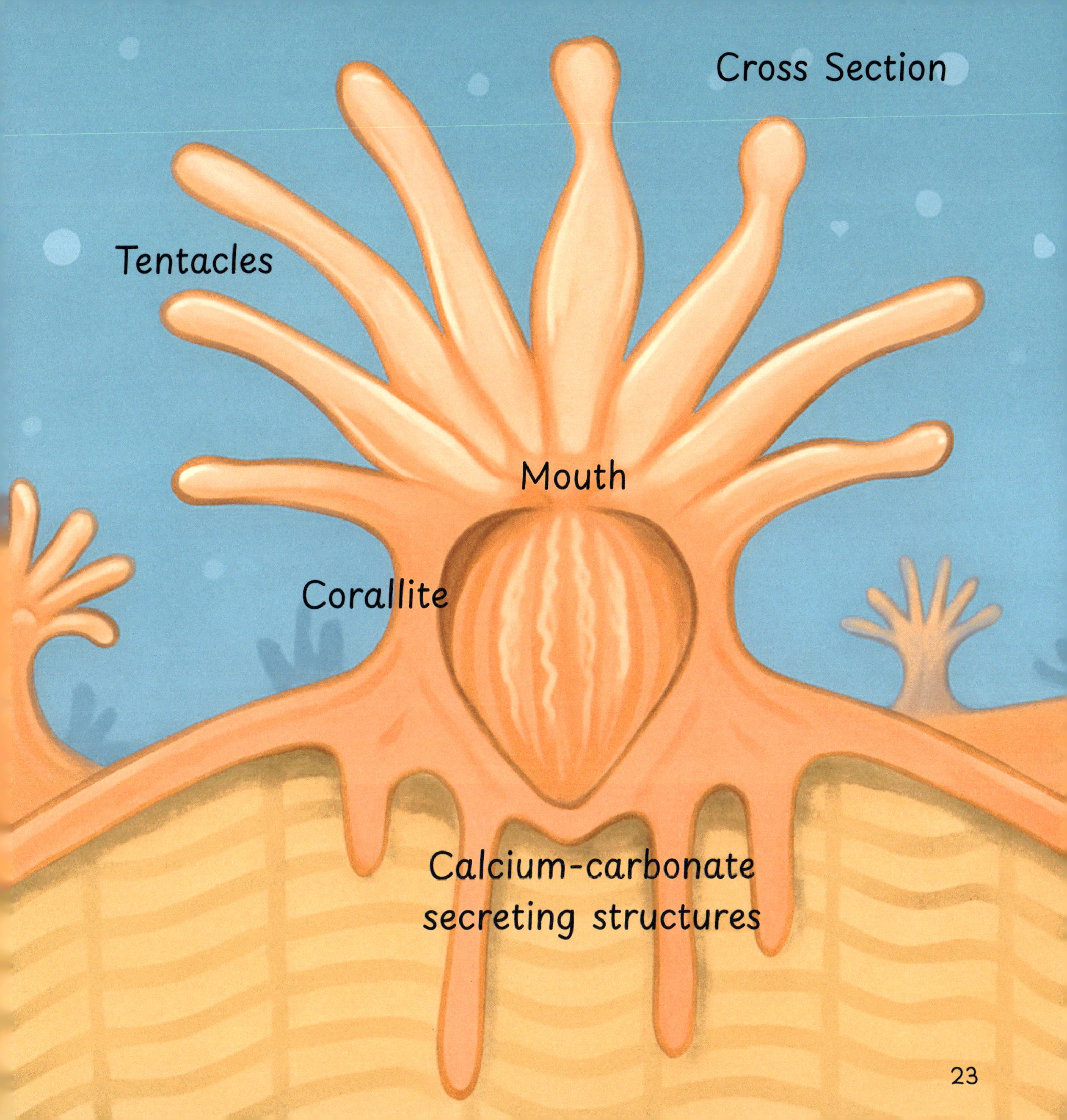
Cross Section
Tentacles
Mouth
Corallite
Calcium-carbonate
secreting structures

Petoskey stones are made mostly of calcite. That means you can polish them until they are super shiny!

Would you like to polish one yourself?
There are directions for you in the back of this book.

Next time you're on a Michigan beach, keep your eyes open. You might just find a tiny piece of ancient sea life waiting for you.

Remember...

Each Petoskey stone is one-of-a-kind, just like you!

Activity Time

Can you find the Petoskey stones?

Each page has two pictures. The top picture shows the stones dry. The bottom shows them wet. When stones get wet, their patterns pop out and are easier to see!

Polishing Your Petoskey Stone

1. **Shape It!**
   a. If needed, gently file your stone to get the shape you want.
2. **Start Sanding (220 grit)**
   a. Wet your stone and sandpaper.
   b. Rub the stone with sandpaper in circles with steady hands.
   c. Rinse and dry. Look for scratches and sand again if needed!
3. **Smooth It More (400 grit)**
   a. Repeat the same motion.
   b. Rinse, dry, and check, no white spots or rough patches!
4. **Final Sanding (600 grit)**
   a. Make it super smooth!
   b. Rinse, dry, and check again. Still see scratches? Go back a step!
5. **Time to Polish!**
   a. Dab a little polishing powder on your damp velvet or corduroy.
   b. Rub the stone in small circles, shine it up!
   c. Still see scratches? Go back to 400 grit & repeat.
6. **Finishing Touch!❤**
   a. Wipe it clean with a soft cloth.
   b. Ta-da! You've got a shiny Petoskey stone for your collection!

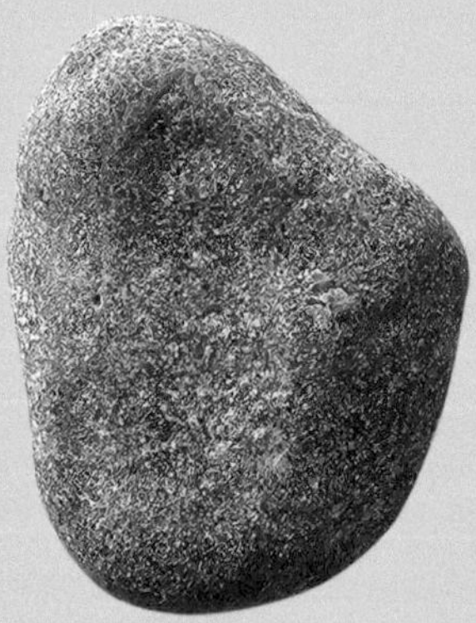

Before

After

Before

After

More Fun Activities

- Heart ♥ Hunt Challenge. Just like Petoskey stones, some hearts are easy to spot and others are camouflaged! A tiny heart hides on every page—even the covers. Finding them will give you great practice in focusing your eyes for your next Petoskey stone hunt!
- Create an ocean scene using paper, fabric, or toys to show the coral's underwater home.
- Go on a rock hunt at a beach or park and look for fossil-like patterns.
- Draw coral polyps and stones based on illustrations in this book.
- Make a timeline to visualize how long ago coral lived.
- Build coral patterns using hexagon shapes from paper or blocks.
- Act out a coral's life with storytelling and imaginative play.
- Paint a Petoskey stone craft using rocks and white paint pens.
- Play a word game with terms like polyp, reef, and fossil.
- Read aloud together using expressive voices or flashlight storytelling.
- Demonstrate sunlight in water with a flashlight and a bowl to show coral's warm habitat.
- Look up real fossil images and compare them to the illustrations.
- Discuss reflection questions, like what it would be like to live in a coral reef.

Explore the Wonders of Petoskey Stones

The links below offer everything you need to explore them, from where to find and identify them, to state collection rules and polishing guides. Whether you're a beginner or enthusiast, these trusted resources will help you get started.

- Geology In. (2017, June). Petoskey Stone: What It Is & Where to Find It. https://www.geologyin.com/2017/06/what-is-petoskey-stone-and-where-can.html
- Doran, A. (2022, February 28). Petoskey Stone. Sisyphos Rocks. https://sisyphos.rocks/petoskey-stone/
- Kchodl, J. (2021, September). Petoskey stones: Where & how to find them. Rock & Gem Magazine. https://www.rockngem.com/what-are-petoskey-stones/
- Mackinac Center for Public Policy. (2011, March 16). 'Petoskey Stone': Michigan's state stone. https://www.mackinac.org/15583
- Michigan Department of Environment, Great Lakes, and Energy. (n.d.). Rock and mineral identification. https://www.michigan.gov/egle/public/learn/geology/rockhounding
- Michigan Department of Environment, Great Lakes, and Energy. (n.d.). The Petoskey Stone. https://www.michigan.gov/-/media/Project/Websites/egle/Documents/Programs/GRMD/Catalog/04/GIMDL-GGPS.PDF
- Michigan Mineralogical Society. (n.d.). Finding state rocks. https://www.michmin.org/finding-state-rocks
- Michigan Rocks. (2020, August 2). How to tumble Petoskey stones [Video]. YouTube. https://youtu.be/grSIZ835VDo

Glossary

Brachiopods: Marine animals with hard shells often found in the same fossil layers as Petoskey stones, providing additional context to the ecosystem of the time.

Calice, Calyx: The cup-shaped depression in the coral skeleton where a polyp resides.

Calcium Carbonate: A chemical compound ($CaCO_3$) secreted by coral polyps to form their skeletons. It is the primary component of coral reefs and Petoskey stones.

Coral: Marine invertebrates within the class Anthozoa. They typically form in colonies and are known for their ability to build reefs through the secretion of calcium carbonate.

Coral Colonies: Groups of coral polyps that live together, forming massive reef structures. These colonies are connected by their calcium carbonate skeletons.

Coral Polyp: The individual, tubular organisms that make up a coral colony. Each polyp has a mouth surrounded by tentacles for feeding.

Corallite: The hard, skeletal cup formed and secreted by an individual coral polyp, serving both as its living chamber and a protective retreat. This structure is often preserved in fossilized coral and provides key evidence of ancient marine life.

Devonian Coral Reef: Ancient coral ecosystems that thrived during the Devonian Period, often preserved as limestone and fossilized coral today.

Devonian Period: A geologic time period approximately 419.2 to 358.9 million years ago, often referred to as the "Age of Fishes." This period saw the flourishing of coral reefs, including Hexagonaria.

Equator: The imaginary line around the Earth that is equidistant from the poles. During the Devonian Period, the area where modern-day Michigan is located was near the equator, providing the warm, shallow seas ideal for coral growth.

Fossilization: The process through which organic materials are replaced by minerals over millions of years, preserving the shape and structure of the original organism.

Hexagonaria: A genus of extinct coral commonly known as Petoskey stones. These fossils are characterized by their hexagonal patterns and were prominent during the Devonian Period.

Limestone: A sedimentary rock composed primarily of calcium carbonate, often formed from accumulated coral and shell debris.

Paleontology: The scientific study of fossils, including coral fossils like Petoskey stones, to understand ancient life and environments.

Petoskey Stone: A type of fossilized coral, specifically Hexagonaria, found primarily in Michigan. These stones display a distinct hexagonal pattern and are often polished for decorative purposes.

Polishing: The process of smoothing and shining Petoskey stones to enhance their natural patterns, making them popular for jewelry and decoration.

Polyps: Small, soft-bodied organisms that form the living part of corals. Each polyp secretes calcium carbonate to build a protective skeleton.

Reef: A structure built by coral colonies, often forming habitats for diverse marine life. Fossilized reefs provide insights into ancient ecosystems.

Sedimentation: The process by which particles settle and accumulate in layers, often covering and preserving coral skeletons, leading to fossilization.

Septum: Radiating vertical support plates within the calyx of the coral skeleton that provide structural integrity and support to the coral polyp.

Tabula: Horizontal skeletal structures within the coral that formed as the polyp deposited calcium carbonate, creating floor-like layers to elevate itself as it grew—especially when buried under sediment. These internal partitions are often visible in fossil cross-sections.

Made in the USA
Monee, IL
12 September 2026

62976610R00024